John Napier and the Invention
of Logarithms, 1614

1616

I · N

MATH 1616

ÆTAIS. 6.6.

Ihone neper of merchistun

John Napier and the Invention of Logarithms, 1614

A LECTURE

E. W. HOBSON, Sc.D., LL.D., F.R.S.

SADLEIRIAN PROFESSOR OF PURE MATHEMATICS
FELLOW OF CHRIST'S COLLEGE, CAMBRIDGE

Cambridge:
at the University Press
1914

CAMBRIDGE UNIVERSITY PRESS
Cambridge, New York, Melbourne, Madrid, Cape Town,
Singapore, São Paulo, Delhi, Tokyo, Mexico City

Cambridge University Press
The Edinburgh Building, Cambridge CB2 8RU, UK

Published in the United States of America by Cambridge University Press, New York

www.cambridge.org
Information on this title: www.cambridge.org/9781107624504

© Cambridge University Press 1914

First published 1914
First paperback edition 2011

A catalogue record for this publication is available from the British Library

ISBN 978-1-107-62450-4 Paperback

JOHN NAPIER AND THE INVENTION OF LOGARITHMS, 1614

In the present year there will be held a celebration, under the auspices of the Royal Society of Edinburgh, of the tercentenary of one of the great events in the history of Science, the publication of John Napier's "Mirifici Logarithmorum Canonis Descriptio," a work which embodies one of the very greatest scientific discoveries that the world has seen. The invention of Logarithms not only marks an advance of the first importance in Mathematical Science, but as providing a great labour-saving instrument for the use of all those who have occasion to carry out extensive numerical calculations it can be compared in importance only with the great Indian invention of our system of numeration.

It is almost always extremely instructive to study in detail the form in which a great discovery or invention was presented by its originator, and to trace in detail the mode in which the fundamental ideas connected with the discovery shaped themselves in his mind, even when, and just because, later developments or simplifications may have so transformed the underlying principles, and still more the practice, of the invention, that we have become accustomed to look at the matter from a point of view, at least superficially, very different from the original one of the discoverer. The case of logarithms is very far from being an exception to this rule; accordingly I propose to give an account, as concise as may be, of the conception of a logarithm in the mind of Napier, and of the methods by which he actually constructed his table of logarithms.

In order fully to appreciate the nature of the difficulties of the task accomplished by the genius of John Napier, some effort of imagination is required, to be expended in realizing the narrowness of the means available in the early part of the seventeenth century for the calculation of the tables, at a time, before the invention of the Differential and Integral Calculus, when calculation by means of infinite series had not yet been invented. Napier's conception of a logarithm involved a perfectly clear apprehension of the nature and consequences of a certain functional relationship, at a time when no general conception of such a relationship had been formulated, or existed in the minds of Mathematicians, and before the intuitional aspect of that relationship had been clarified by means of the great invention of coordinate geometry made later in the century by René Descartes. A modern Mathematician regards the logarithmic function as the inverse of an exponential function ; and it may seem to us, familiar as we all are with the use of operations involving indices, that the conception of a logarithm would present itself in that connection as a

fairly obvious one.   We must however remember
that, at the time of Napier, the notion of an index,
in its generality, was no part of the stock of ideas
of a Mathematician, and that the exponential
notation was not yet in use.

Summary of the life of Napier.

I must content myself with giving an exceed-
ingly brief account of the external facts of the life
of Napier*.

John Napier†, the eighth Napier of Merchiston,
usually described as Baron, or Fear, of Merchiston,
was born at Merchiston near Edinburgh in 1550,
when his father Archibald Napier was little more
than sixteen years old.   John Napier matriculated
at St Andrews in 1563, but did not stay there

---

* For a full account of the life and activities of Napier the
"Memoirs of John Napier of Merchiston" by Mark Napier,
published in 1834, may be consulted.

† The name Napier was spelled in various ways, several
of which were used by John Napier; thus we find Napeir,
Nepair, Nepeir, Neper, Nepper, Naper, Napare, Naipper.

sufficiently long to graduate, as he departed
previous to 1566 in order to pursue his studies
on the Continent, whence he returned to Merchiston in or before 1571. His first marriage, by
which he had one son Archibald who was raised
to the peerage in 1627 as Lord Napier, and one
daughter, took place in 1572. A few years after
the death of his wife in 1579, he married again.
By his second marriage he had five sons and
five daughters; the second son, Robert, was his
literary executor. The invasion of the Spanish
Armada in 1588 led Napier, as an ardent Protes-
tant, to take a considerable part in Church politics.
In January 159$\frac{3}{4}$ he published his first work "A
plaine discovery of the whole Revelation of
St John." This book is regarded as of considerable importance in the history of Scottish theo-
logical literature, as it contained a method of
interpretation much in advance of the age; it
passed through several editions in English, French,
German, and Dutch.

In July 1594, Napier entered into a curious contract with a turbulent baron, Robert Logan of Restalrig, who had just been outlawed. In this contract, which appears to shew that John Napier was not free from the prevalent belief in Magic, he agreed to endeavour to discover a treasure supposed to lie hidden in Logan's dwelling-place, Fast Castle. Napier was to receive a third part of the treasure when found, in consideration that "the said Jhone sall do his utter & exact diligens to serche & sik out, and be al craft & ingyne that he dow, to tempt, trye, and find out the sam, and be the grace of God, ather sall find the sam, or than mak it suir that na sik thing hes been thair; sa far as his utter trawell diligens and ingyne may reach."

In a document dated June 7, 1596, Napier gave an account of some secret inventions he had made which were "proffitabill & necessary in theis dayes for the defence of this Iland & withstanding of strangers enemies of God's truth &

relegion." His activities in this direction were no doubt stimulated by the fear of the generally expected invasion by Philip of Spain. It is interesting to note them, in view of the military tastes of many of his descendants. The inventions consisted of a mirror for burning the enemies' ships at any distance, of a piece of artillery capable of destroying everything round an arc of a circle, and of a round metal chariot so constructed that its occupants could move it rapidly and easily, while firing out through small holes in it. Napier's practical bent of mind was also exhibited in the attention he paid to agriculture, especially on the Merchiston estate, where the land was tilled by a system of manuring with salt.

There is evidence that Mathematics occupied Napier's attention from an early age. From a MS. that was first published in 1839 under the title "De Arte Logistica" it appears that his investigations in Arithmetic and Algebra had led him to a consideration of the imaginary roots of

equations, and to a general method for the extraction of roots of numbers of all degrees.   But,
led probably by the circumstances of the time, he
put aside this work in order to devote himself to
the discovery of means of diminishing the labour
involved in numerical computations.   The second
half of the sixteenth century was the time in which
the Mathematicians of the Continent devoted a
great deal of attention to the calculation of tables
of natural trigonometrical functions.   The most
prominent name in this connection is that of
Georg Joachim Rheticus, the great computer
whose work has never been superseded, and the
final result of whose labours is embodied in the
table of natural sines for every ten seconds to
fifteen places of decimals, published by Pitiscus in
1613, the year before the publication by Napier
of the discovery which was destined to revolu-
tionize all the methods of computation, and to
substitute the use of logarithmic for that of natural
trigonometrical functions.   It was in the early

years of the seventeenth century that Johannes
Kepler was engaged in the prodigious task of
discovering, and verifying by numerical calculation,
the laws of the motion of the planets. In this age
of numerical calculation then Napier occupied
himself with the invention of methods for the
diminution of the labour therein involved. He
himself states in his "Rabdologia," to which
reference will presently be made, that the
canon of logarithms is "a me longo tempore
elaboratum." It appears from a letter of Kepler
that a Scotsman, probably Thomas Craig, a
friend of the Napier family, gave the astronomer
Tycho Brahe in the year 1594 hopes that an
important simplification in the processes of arithmetic would become available. There is strong
evidence that Napier communicated his hopes to
Craig twenty years before the publication of the
Canon.

The "Descriptio," of which an account will
be given presently, was as stated at the outset

published in 1614. About the same time Napier devised several mechanical aids for the performance of multiplications and divisions and for the extraction of square and cube roots. He published an account of these inventions in 1617 in his " Rabdologia," as he says, " for the sake of those who may prefer to work with the natural numbers." The method which Napier calls Rabdologia consists of calculation of multiplications and divisions by means of a set of rods, usually called "Napier's bones." In 1617, immediately after the publication of the " Rabdologia," Napier died.

The " Descriptio" did not contain an account of the methods by which the "wonderful canon " was constructed. In an "Admonitio" printed at the end of Chapter II, Napier explains that he prefers, before publishing the method of construction, to await the opinion of the learned world on the canon ; he says " For I expect the judgement & censure of learned men hereupon, before the rest rashly published, be exposed to the detraction of the envious."

The "Mirifici Logarithmorum Canonis Constructio," which contains a full explanation of the method of construction of the wonderful canon, and a clear account of Napier's theory of logarithms, was published by his son Robert Napier in 1619. In the preface by Robert Napier it is stated that this work was written by his father several years before the word "logarithm" was invented, and consequently at an earlier date than that of the publication of the "Descriptio." In the latter the word "logarithm" is used throughout, but in the "Constructio," except in the title, logarithms are called "numeri artificiales." After explaining that the author had not put the finishing touch to the little treatise, the Editor writes "Nor do I doubt that this posthumous work would have seen the light in a much more perfect & finished state, if God had granted a longer enjoyment of life to the Author, my most dearly

beloved father, in whom, by the opinion of the
wisest men, among other illustrious gifts this
shewed itself pre-eminent, that the most difficult
matters were unravelled by a sure and easy
method, as well as in the fewest words."

Reception of the Canon by Contemporary Mathematicians.

The new invention attracted the attention
of British and Foreign Mathematicians with a
rapidity which may well surprise us when we take
into account the circumstances of the time. In
particular the publication of the wonderful canon
was received by Kepler with marked enthusiasm.
In his "Ephemeris" for 1620, Kepler published
as the dedication a letter addressed to Napier,
dated July 28, 1619, congratulating him warmly
on his invention and on the benefit he had con-
ferred upon Astronomy. Kepler explains how he
verified the canon and found no essential errors

in it, beyond a few inaccuracies near the beginning of the quadrant. The letter was written two years after Napier's death, of which Kepler had not heard. In 1624 Kepler himself published a table of Napierian logarithms with modifications and additions. The "Descriptio," on its publication in 1614, at once attracted the attention of Henry Briggs (1556–1630), Fellow of St John's College, Cambridge, Gresham Professor of Geometry in the City of London, and afterwards Savilian Professor of Geometry at Oxford, to whose work as the successor of Napier in the task of construction of logarithmic tables in an improved form I shall have to refer later. In a letter to Archbishop Ussher, dated Gresham House, March 10, 1615, Briggs wrote, "Napper, lord of Markinston, hath set my head & hands a work with his new & admirable logarithms. I hope to see him this summer, if it please God, for I never saw book which pleased me better, or made me more wonder." Briggs visited Napier, and stayed with him a month in 1615, again visited him in 1616, and intended to visit him again in 1617, had Napier's life been spared. Another eminent English Mathematician, Edward Wright, a Fellow of Gonville and Caius College, who at once saw the importance of logarithms in connection with navigation, in the history of which he occupies a conspicuous place, translated the "Descriptio," but died in 1615 before it could be published. The translation was however published in 1618 by his son Samuel Wright.

The contents of the "Descriptio" and of the "Constructio."

The "Descriptio" consists of an ornamental title page, fifty-seven pages of explanatory matter, and ninety pages of tables. A specimen page of the tables is here reproduced. The explanatory matter contains an account of Napier's conception of a logarithm, and of the principal properties of logarithms, and also of their application in the

9 min	Sinus	Logarithmi	Differentiæ	logarithmi	Sinus	
0	1564345	18551174	18427293	123881	9876883	60
1	1567218	18532826	18408484	124342	9876427	59
2	1570091	18514511	18389707	124804	9875971	58
3	1572964	18496231	18370964	125267	9875514	57
4	1575837	18477984	18352253	125731	9875056	56
5	1578709	18459772	18333576	126196	9874597	55
6	1581581	18441594	18314933	126661	9874137	54
7	1584453	18423451	18296324	127127	9873677	53
8	1587325	18405341	18277747	127594	9873216	52
9	1590197	18387265	18259203	128062	9872754	51
10	1593069	18369223	18240692	128531	9872291	50
11	1595941	18351214	18222213	129001	9871827	49
12	1598812	18333237	18203765	129472	9871362	48
13	1601684	18315294	18185351	129943	9870897	47
14	1604555	18297384	18166969	130415	9870431	46
15	1607426	18279507	18148619	130888	9869964	45
16	1610297	18261663	18130301	131362	9869496	44
17	1613168	18243851	18112014	131837	9869027	43
18	1616038	18226071	18093758	132313	9868557	42
19	1618909	18208323	18075533	132790	9868087	41
20	1621779	18190606	18057338	133268	9867616	40
21	1624649	18172924	18039177	133747	9867144	39
22	1627519	18155273	18021047	134226	9866671	38
23	1630389	18137654	18002948	134706	9866197	37
24	1633259	18120067	17984880	135187	9865722	36
25	1636129	18102511	17966842	135669	9865246	35
26	1638999	18084987	17948835	136152	9864770	34
27	1641868	18067495	17930859	136636	9864293	33
28	1644738	18050034	17912913	137121	9863815	32
29	1647607	18032604	17894997	137607	9863336	31
30	1650476	18015207	17877114	138093	9862856	30 min Gr. 80

solution of plane and spherical triangles. Napier's well-known rules of circular parts containing the complete system of formulae for the solution of right-angled spherical triangles are here given. The logarithms given in the tables are those of the sines of angles from o° to 90° at intervals of one minute, to seven or eight figures. The table is arranged semi-quadrantally, so that the logarithms of the sine and the cosine of an angle appear on the same line, their difference being given in the table of differentiae which thus forms a table of logarithmic tangents. It must be remembered that, at that time and long afterwards, the sine of an angle was not regarded, as at present, as a ratio, but as the length of that semi-chord of a circle of given radius which subtends the angle at the centre. Napier took the radius to consist of 10^7 units, and thus the sine of 90°, called the whole sine, is 10^7; the sines of smaller angles decreasing from this value to zero. The table is therefore one of the logarithms of numbers between 10^7 and 0, not for equidistant numbers, but for the numbers corresponding to equidistant angles. It is important to observe that the logarithms in Napier's tables are not what we now know under the name of Napierian or natural logarithms, i.e. logarithms to the base e. His logarithms are more closely related to those to the base $1/e$; the exact relation is that, if x is a number, and $\overset{\text{Nap}}{\text{Log}}\, x$ its logarithm in accordance with Napier's tables, $\dfrac{\overset{\text{Nap}}{\text{Log}}\, x}{10^7}$ is the logarithm of $\dfrac{x}{10^7}$ to the base $1/e$; thus

$$\frac{x}{10^7} = \left(\frac{1}{e}\right)^{\dfrac{\overset{\text{Nap}}{\text{Log}\, x}}{10^7}},$$

or $\qquad \overset{\text{Nap}}{\text{Log}}\, x = 10^7 \log_e 10^7 - 10^7 \log_e x.$

Napier had no explicit knowledge of the existence of the number e, nor of the notion of the base of a system of logarithms, although as we shall see he was fully cognizant of the arbitrary element in the possible systems of logarithms. His choice was made with a view to making the logarithms of the sines of angles between 0° and 90°, i.e. of numbers between 0 and 10⁷, positive and so as to contain a considerable integral part.

The "Constructio" consists of a preface of two pages, and fifty-seven pages of text. The conception of a logarithm is here clearly explained, and a full account is given of the successive steps by which the Canon was actually constructed. In this work one of the four formulae for the solution of spherical triangles, known as Napier's analogies, is given, expressed in words; the other three formulae were afterwards added by Briggs, being easily deducible from Napier's results.

The decimal point.

Our present notation for numbers with the decimal point appears to have been independently invented by Napier, although a point or a half bracket is said to have been employed somewhat

earlier by Jobst Bürgi for the purpose of separat-
ing the decimal places from the integral part of a
number*. The invention of decimal fractions was
due to Simon Stevin (1548–1620), who published
a tract in Dutch, " De Thiende," in 1585, and in
the same year one in French under the title " La
Disme," in which the system of decimal fractions
was introduced, and in which a decimal system of
weights, measures and coinage was recommended.
In the "Rabdologia" Napier refers to Stevin in an
"Admonitio pro Decimali Arithmetica" in which
he emphasizes the simplification arising from the
use of decimals, and introduces the notation with
the decimal point.   By Stevin and others the
notation 94 ⓪ 1 ① 3 ② 0 ③ 5 ④ or 94 1′ 3″ 0‴ 5⁗,
for example, was used instead of Napier's notation
94·1305.   Later on Briggs sometimes used the
notation 94<u>1305</u>.   It is clear that the notation
introduced by Napier, which was however not

* The decimal point was also employed by Pitiscus in the
tables appended to the later edition, published in 1612, of his
"Trigonometria."

universally adopted until the eighteenth century, was far better adapted than the more complicated notation used by Stevin and later writers to bring out the complete parity of the integral and decimal parts of a number in relation to the operations of arithmetic, and to emphasize the fact that the system of decimal fractions involves only an extension of the fundamental conception of our system of notation for integral numbers, that the value of a digit, in relation to the decimal scale, is completely indicated by its position.

Napier's definition of a logarithm.

$$p_1 \quad T \qquad P_1 \quad P_2\, P_3\, P_4 \qquad\qquad S$$

$$q_1 \quad T_1 \qquad\qquad Q_1 \quad Q_2\, Q_3\, Q_4 \qquad \text{to } \infty$$

Napier supposes that on a straight line TS which he takes to consist of 10^7 units, the radius of the circle for which the sines are measured, a point P moves from left to right so that its velocity is at every point proportional to the distance from

S. He supposes that on another straight line a point Q moves with uniform velocity equal to that which P has when at T, and that Q is at T_1, when P is at T. When P has any particular position P_1 in the course of its motion, the logarithm of the sine or length SP_1 is defined to be the number representing the length T_1Q_1, from T_1 to the position Q_1 of Q at the time when P is at P_1. Thus the logarithm of the whole sine $TS\,(=10^7)$ is o, and the logarithm of any sine less than 10^7 is positive and increases indefinitely as the sine diminishes to zero. Napier recognized that in accordance with this definition, the logarithm of a number greater than 10^7, corresponding to Sp_1, where p_1 is the position of P when it has not yet reached T, will be negative, the corresponding position of Q being on the left of T_1.

Let Q_1, Q_2, Q_3, Q_4, ... be a number of positions of Q such that $Q_1Q_2 = Q_2Q_3 = Q_3Q_4 = ...$ and let P_1, P_2, P_3, ... be the corresponding positions of P; so that P_1P_2, P_2P_3, P_3P_4, ... are spaces described by P in equal times. Napier then shews by means of special illustrations that

$$SP_1 : SP_2 = SP_2 : SP_3 = \ldots,$$

and thus that, corresponding to a series of values, of T_1Q that are in arithmetic progression, there are a series of values of SP that are in geometric progression.

The matter may be put in a concise form which represents the gist of Napier's reasoning, and of the essential point of which he had a clear intuition.

Let $\dfrac{P_1S}{P_2S} = \dfrac{P_2S}{P_3S}$; and let p be any point in P_1P_2, and p' the corresponding point in P_2P_3; so that $P_1p : pP_2 = P_2p' : p'P_3$. The velocity of the moving point when at p bears a constant ratio $\left(\dfrac{SP_1}{SP_2} \equiv \dfrac{SP_2}{SP_3} \equiv \dfrac{P_1P_2}{P_2P_3}\right)$ to its velocity when at p'. As this holds for every corresponding pair of points p, p' in the two intervals P_1P_2, P_2P_3, it is clear that the motion in P_1P_2 takes place in the same time as that in P_2P_3; the velocities at all corresponding points being changed in the same ratio, that of P_1P_2 to P_2P_3. Hence the result follows that

$$\frac{P_1S}{P_2S} = \frac{P_2S}{P_3S} = \frac{P_3S}{P_4S} = \cdots,$$

if the points Q_1, Q_2, Q_3, Q_4, ... are such that $Q_1Q_2 = Q_2Q_3 = Q_3Q_4 = \ldots$; i.e. if the spaces P_1P_2, P_2P_3, P_3P_4, ... are described in equal times. Thus the logarithms of a set of numbers in geometric progression are themselves in arithmetic progression.

In our modern notation, if $x = SP$, we have $\frac{dx}{dt} = - \frac{Vx}{10^7}$, where V denotes the velocity of P at T; and if $y = T_1Q$, $\frac{dy}{dt} = V$; thus $\frac{dx}{dy} = - \frac{x}{10^7}$; and accordingly Napier's method amounts to an intuitional representation of the integration of this differential equation.

The limits of a logarithm.

As no method was available by which a logarithm could be calculated to an arbitrarily great degree of approximation, Napier obtained two limits between which a logarithm must lie, and his whole method of construction depends upon the use of these limits, together with corresponding limits for the value of the difference of the logarithms of two numbers.

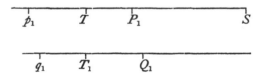

Since the velocities of P and Q at T, T_1 are the same, and the velocity of P decreases afterwards whereas that of Q remains constant, it is clear that $TP_1 < T_1Q_1$. Again let p_1T on the left of T be described in the same time as TP_1, so that $q_1T_1 = T_1Q_1$. It is then clear that $p_1T > q_1T_1$.

If $x = P_1S$, $\overset{\text{Nap}}{\text{Log}} x = T_1Q_1$, we thus have

$$\overset{\text{Nap}}{\text{Log}} x > TP_1, \quad \text{or} \quad 10^7 - x;$$

and $\overset{\text{Nap}}{\text{Log}} x = q_1 T_1 < p_1 T$, or $TP_1 \dfrac{10^7}{x}$, which is $(10^7 - x) \dfrac{10^7}{x}$. Thus

$$(10^7 - x) \frac{10^7}{x} > \overset{\text{Nap}}{\text{Log}} x > 10^7 - x \quad \ldots (1);$$

these are Napier's limits for a logarithm.

In a similar manner it is shewn that

$$10^7 \frac{y - x}{x} > \overset{\text{Nap}}{\text{Log}} x - \overset{\text{Nap}}{\text{Log}} y > 10^7 \frac{y - x}{y} \ldots (2),$$

where $x < y$; these are the limits which Napier employs for $\overset{\text{Nap}}{\text{Log}} x - \overset{\text{Nap}}{\text{Log}} y$.

The results (1) and (2) were given by Napier in words, no short designation being employed even for a logarithm.

Napier's construction of the canon.

The first step taken by Napier in the process of constructing the canon was to form three tables of numbers in geometric progression. The first table consists of 101 numbers, of which 10^7 is the first, and of which $1 - \dfrac{1}{10^7}$ is the common ratio; thus (in modern notation) the table consists of the numbers $10^7 \left(1 - \dfrac{1}{10^7}\right)^r$, where r has the values 0 to 100. Each number was formed by subtracting from the preceding one the number obtained by moving the digits seven places to the right.

First Table

$$\left\{ 10^7 \left(1 - \frac{1}{10^7}\right)^r,\ r = 0 \text{ to } 100 \right\}$$

```
10000000·0000000
      1·0000000
 ───────────────
 9999999·0000000
       ·9999999
 ───────────────
 9999998·0000001
       ·9999998
 ───────────────
 9999997·0000003
       ·9999997
 ───────────────
 9999996·0000006
```

to be continued up to

9999900·0004950

The second table consists of the 51 numbers $10^7 \left(1 - \dfrac{1}{10^5}\right)^r$, where $r = 0, 1, \ldots 50$. The common ratio $1 - \dfrac{1}{10^5}$ is nearly equal to $\left(1 - \dfrac{1}{10^7}\right)^{100}$, that of the last number in the first table, to the first in that table.

Second Table

$$\left\{10^7 \left(1 - \dfrac{1}{10^5}\right)^r, \ r = 0 \text{ to } 50\right\}$$

$$
\begin{array}{r}
10000000{\cdot}000000 \\
100{\cdot}000000 \\
\hline
9999900{\cdot}000000 \\
99{\cdot}999000 \\
\hline
9999800{\cdot}001000
\end{array}
$$

to be continued up to

$$9995001{\cdot}222927$$

In this table there is an arithmetical error, as the last number should be $9995001{\cdot}224804$; the effect of this error on the canon will be referred to later.

The ratio of the last number to the first is $\left(1 - \dfrac{1}{10^5}\right)^{50}$, which is nearly $1 - \dfrac{1}{2000}$.

The third table consists of 69 columns, and each column contains 21 numbers. The first number in any column is obtained by taking $1 - \frac{1}{100}$ of the first number in the preceding column. The numbers in any one column are obtained by successive multiplication by $1 - \frac{1}{2000}$; thus the pth number in the qth column is

$$10^7 \left(1 - \tfrac{1}{2000}\right)^{p-1} \left(1 - \tfrac{1}{100}\right)^{q-1}.$$

Third Table

First column	Second column		69th column
10000000·0000	9900000·0000	...	5048858·8900
9995000·0000	9895050·0000	...	5046334·4605
9990002·5000	9890102·4750	...	5043811·2932
9985007·4987	9885157·4237	...	5041289·3879
continued to	continued to		continued to
9900473·5780	9801468·8423	...	4998609·4034

The ratio of the last to the first number in any one column is $\left(1 - \frac{1}{2000}\right)^{20}$, which is nearly $1 - \frac{1}{100}$ or $\frac{99}{100}$.

It will be observed that the last number of the last column is less than half the radius, and thus corresponds to the sine of an angle somewhat less than 30°.

In this table there are, speaking roughly, 68 numbers in the ratio 100 : 99 interpolated between 10⁷ and ½10⁷; and between each of these are interpolated twenty numbers in the ratio 10000 : 9995.

Having formed these tables, Napier proceeds to obtain with sufficient approximation the logarithms of the numbers in the tables. For this purpose his theorems (1) and (2) as to the limits of logarithms are sufficient. In the first table, the logarithm of 9999999 is, in accordance with (1), between 1·0000001 and 1·0000000, and Napier takes the arithmetic mean 1·00000005 for the required logarithm. The logarithm of the next sine in the table is between 2·0000002 and 2·0000000; for this he takes 2·00000010, for the next sine 3·00000015, and so on.

The theorem (2) is used to obtain limits for the logarithms of numbers nearly equal to a number in the first table. In this way the logarithm of 9999900 the second number in the second table is found to lie between 100·0005050 and 100·0004950; the next logarithm has limits double of these, and so on. The logarithm of the last sine in the second table is thus found to lie between 5000·0252500 and 5000·0247500. The logarithms of the numbers in the second table having thus been found to a sufficient degree of approximation, the logarithm of a number near one in the second table is found thus :—Let y be the given sine, x the nearest sine in the table ; say $y < x$. Determine z so that $\dfrac{z}{10^7} = \dfrac{y}{x}$, then

$$\overset{\text{Nap}}{\text{Log}} z = \overset{\text{Nap}}{\text{Log}} z - \overset{\text{Nap}}{\text{Log}} 10^7 = \overset{\text{Nap}}{\text{Log}} y - \overset{\text{Nap}}{\text{Log}} x \; ;$$

find the limits of $\overset{\text{Nap}}{\text{Log}} z$ by means of the first table, and when these are found add them to those of $\overset{\text{Nap}}{\text{Log}} x$, and we thus get the limits of $\overset{\text{Nap}}{\text{Log}} y$. In this manner limits are found for the logarithms of all the numbers in the first column of the third table; thus those of 9900473·57808 are 100024·9657720 and 100024·9757760, and the

logarithm is taken to be 100024·9707740, the
mean of the two limits. The first number in
the second column differs only in the fifth cypher
from the last number in the first column, and thus
its logarithm can be calculated approximately.
The logarithms of all the other numbers in the
table can then be found, since the logarithms of
all the numbers in any one column, or in any one
row, are in arithmetic progression.

When the logarithms of all the numbers in
the third table have thus been calculated, the
table formed by filling them in is called by
Napier his radical table, and is of the form
given on the opposite page.

The radical table being completed, the logarithms in it are employed for the calculation of
the principal table or canon. For this purpose
the logarithms of sines very nearly equal to the
whole sine 10^7 are obtained simply by subtracting
the given sine from 10^7. The logarithm of a sine
embraced within the limits of the radical table is

The Radical Table

First column		Second column		69th column	
Natural numbers	Logarithms	Natural numbers	Logarithms	Natural numbers	Logarithms
10000000·0000	·0	9900000·0000	100503·3	5048858·8900	6834225·8
9995000·0000	5·0012	9895050·0000	105504·6	5046334·4605	6839227·1
9990002·5000	10·0025	9890102·4750	110505·8	5043811·2932	6844228·3
9985007·4987	15·0037	9885157·4237	115507·1	5041289·3879	6849229·6
9900473·5780	100025·0	9801468·8423	200528·2	4998609·4034	6934250·8

found thus:—Let x be the given sine, y the nearest sine in the table; say $x > y$; calculate $(x-y)\,10^7$ and divide it either by x or by y, or by some number between the two; then add the result to the logarithm of the table sine.

For the purpose of finding the logarithm of a sine which is not embraced within the limits of the radical table, Napier gives a short table containing the difference of the logarithms of two sines of which the ratio is compounded of the ratios $2:1$, and $10:1$.

Short Table

Given proportion of sines	Corresponding difference of logarithms	Given proportion of sines	Corresponding difference of logarithms
2 to 1	6931469·22	8000 to 1	89871934·68
4 ,,	13862938·44	10000 ,,	92103369·36
8 ,,	20794407·66	20000 ,,	99034838·58
10 ,,	23025842·34	40000 ,,	105966307·80
20 ,,	29957311·56	80000 ,,	112897777·02
40 ,,	36888780·78	100000 ,,	115129211 70
80 ,,	43820250·00	200000 ,,	122060680·92
100 ,,	46051684·68	400000 ,,	128992150·14
200 ,,	52983153·90	800000 ,,	135923619 36
400 ,,	59914623·12	1000000 ,,	138155054·04
800 ,,	66846092·34	2000000 ,,	145086523·26
1000 ,,	69077527·02	4000000 ,,	152017992·48
2000 ,,	76008996·24	8000000 ,,	158949461·70
4000 ,,	82940465·46	10000000 ,,	161180896·38

To calculate this table, Napier found $\overset{\text{Nap}}{\text{Log}}$ 10⁷ and $\overset{\text{Nap}}{\text{Log}}$ 500000 by using the radical table; and thus 6931469·22 was found as the difference of the logarithms of numbers in the ratio 2 : 1. The difference of logarithms of sines in the ratio 8 : 1 is three times 6931469·22, i.e. 20794407·66. The sine 8000000 is found by using the radical table to have for its logarithm 2231434·68, whence by addition the logarithm of the sine 1000000 is found to be 23025842·34. Since the radius is ten times this sine, all sines in the ratio 10 : 1 will have this number for the difference of their logarithms. The rest of the table was then calculated from these determinations.

The logarithms of all sines that are outside the limits of the radical table could now be determined. Multiply the given sine by 2, 4, 8, ... 200, ... or by any proportional number in the short table, until a number within the limits of the radical table is found. Find the logarithm of the sine given by the radical table, and add to it the difference which the short table indicates.

In the manner described, the logarithms of all sines of angles between 0° and 45° could now be determined, and the principal table or canon completed. Napier gave, however, a rule by which when the logarithms for all the angles not less than 45° are known, the logarithms for all the angles less than 45° can be determined. This rule we may write in the form

$$\overset{\text{Nap}}{\text{Log}} \tfrac{1}{2} 10^7 + \overset{\text{Nap}}{\text{Log}} \sin x$$

$$= \overset{\text{Nap}}{\text{Log}} \sin \tfrac{1}{2} x + \overset{\text{Nap}}{\text{Log}} \sin (90° - \tfrac{1}{2} x),$$

which follows from the fact that $\dfrac{\cos \tfrac{1}{2} x}{\sin x} = \dfrac{\tfrac{1}{2}}{\sin \tfrac{1}{2} x}$ when we take into account the change in the definition of a sine.

The accuracy of Napier's Canon.

It has been observed above that a numerical error occurs in the value of the last number in the second table. As Napier employed this

inaccurate value in his further calculations, it
produced an error in the greater part of his
logarithmic tables. The effect of this error is
that most of the logarithms are diminished by
about $\frac{3}{8} 10^{-6}$ of their correct values. Napier
himself observes in the " Constructio " that some
of the logarithms he obtained by means of his rule
for finding the logarithms of numbers outside the
limits of the radical table differ in value from
the logarithms of the same numbers found by the
rule for determining the logarithms of sines of
angles less than $45°$. He attributes this dis-
crepancy to defects in the values of the natural
sines he employed, and suggested a recalculation
of natural sines in which 10^8 should be the radius.
Owing to these two causes, the last figure in the
logarithms of the canon is not always correct.

The improved system of logarithms.

The special purpose of application to trigonometrical calculations accounts for Napier's choice

of the system in which the logarithm of 10^7 is zero, and the logarithms of sines of angles between $0°$ and $90°$ are positive. It is, however, clear that the rule of the equality of the sum of the logarithms of two numbers and that of their product would hold for numbers in general, only if the logarithm of unity were taken to be zero, as a number is unaltered by multiplication by unity. On this account, Napier, in an appendix to the "Constructio," proposed the calculation of a system of logarithms in which Log $1 = 0$, and Log $10 = 10^{10}$. This is practically equivalent to the assumption Log $10 = 1$, as the former assumption merely indicates that the logarithms are to be calculated to 10 places of decimals. Briggs pointed out, in his lectures at Gresham College, that a system would be convenient, on which 0 should be the logarithm of 1, and 10^{10} that of the tenth part of the whole sine (viz. sin $5° 44' 21''$), which would be equivalent to Log $\frac{1}{10} = 10^{10}$. This system he suggested to Napier during his visit to Merchiston in 1615, when Napier pointed out that the same idea had occurred to himself, but that the assumption Log 10 = 10¹⁰ would lead to the most convenient system of all, and this was at once admitted by Briggs.

In the appendix above referred to, Napier gives some indications of methods by which the improved logarithms might be calculated. These depend upon exceeding laborious successive extractions of fifth and of square roots, which work he proposed should be carried out by others, and especially by Briggs. In an "Admonitio" printed in the "Constructio," Napier remarked that it is a matter of free choice to what sine or number the logarithm o is assigned, that it is necessary frequently to multiply or divide by the complete sine (sin 90°), and thus that a saving of trouble arises if the logarithm of this sine be taken to be zero.

Briggs immediately set about the calculation of these improved logarithms, and in the following year, when he again visited Napier, shewed him a large part of the table which was afterwards published in 1624. On the death of Napier in 1617 the whole work of developing the new invention passed into the skilful hands of Briggs, who, in the same year, published his " Logarithmorum Chilias Prima," containing the common or Briggian logarithms of the first thousand numbers to 14 places of decimals. In 1624 he published the "Arithmetica Logarithmica," a table of logarithms of the first 20000 numbers and of the numbers from 90000 to 100000, to 14 places of decimals. The gap between 20000 and 90000 was fitted up by Adrian Vlacq, who published in 1628 at Gouda a table of common logarithms of numbers from 1 to 100000, to 10 places of decimals. Vlacq's tables, although not free from error, have formed the basis of all the numerous tables of logarithms of natural numbers that have been since published.

Other Tables.

A table of logarithms exactly similar to those of Napier in the "Constructio" was published in 1624 by Benjamin Ursinus at Cologne. The intervals of the angles are 10″, and the logarithms are given to 8 places. The first logarithms to the base e were published by John Speidell in his "New Logarithmes," in London in 1619; this table contains logarithmic sines, tangents and secants for every minute of the quadrant to 5 decimal places.

Predecessors of Napier.

It is usually the case that the fundamental conceptions involved in a great new invention have a history, which reaches back to a time, often a long time, before that of the inventor. Although Napier's introduction of logarithms is justly entitled to be regarded as a really new invention, it is not an exception to the usual rule. The notion of an integral power of a ratio was employed by the Greek Mathematicians. The nature of the correspondence between a geometric progression and an arithmetic progression had been observed by various Mathematicians. In particular Michael Stifel (1486–1567), in his celebrated " Arithmetica Integra," published in 1544, expressly indicated the relations between operations with the terms of a geometric and an arithmetic series, in which the terms are made to correspond, viz. the relations between multiplication, division and exponentiation on the one hand, and addition, subtraction and multiplication or division by an integer on the other hand. But no indication was given by Stifel or others how this correspondence could be utilized for the purpose of carrying out difficult arithmetical calculations. There were even given by the Belgian Mathematician Simon Stevin (1548–1620) certain special tables for the calculation of interest, consisting of tables of the values of

$$\frac{1}{(1+r)^n}, \text{ and of } \frac{1}{1+r} + \frac{1}{(1+r)^2} + \cdots + \frac{1}{(1+r)^n}.$$

The first of these tables are really tables of antilogarithms, but there were given no theoretical explanations which would extend the use of the tables beyond their special purpose. Napier, whether he was acquainted with Stifel's work or not, was the first whose insight enabled him to develop the theoretical relations between geometric and arithmetic series into a method of the most far-reaching importance in regard to arithmetic calculations in general. On the theoretical side, Napier's representation by continuously moving points involved the conception of a functional relationship between two continuous variables, whereas Stifel and others had merely considered the relationship between two discrete sets of numbers. This was in itself a step of the greatest importance in the development of Mathematical Analysis.

A rival inventor.

No account of the invention of logarithms would be complete without some reference to the work of Jobst Bürgi (1552–1632), a Swiss watch-maker and instrument-maker, who independently invented a system of logarithms. His system was published in 1620, after Napier's Canon had become known and fully recognized, in a work entitled "Arithmetische und Geometrische Progress-Tabulen." The table is really an antilogarithmic table, and consists of a set of numbers printed red placed in correspondence with a set of numbers printed black. The red numbers are 0, 10, 20, ... , those of an arithmetic series, and the corresponding black numbers are 100000000, 100010000, 100020001, of a geometric series ; thus the red numbers are the logarithms of the black ones divided by 10^8 with the base $\sqrt[10]{1\cdot0001}$. Bürgi appears to have devised his system a good many years before he published it, but kept it secret until he published his tables six years after the appearance of those of Napier.

Conclusion.

The system of Bürgi is decidedly inferior to that of Napier, and the knowledge of the use of logarithms which was spread in the scientific world was entirely due to the work of Napier.

The concensus of opinion among men of Science of all nations has ascribed to Napier the full honour due to the inventor of the method which has provided the modern world with a tool that is indispensable for all elaborate arithmetical calculations. In the great advance which had taken place in Mathematical Science during the half century preceding the publication of the "Constructio," British Mathematicians had taken no part. It is very remarkable that, in a country distracted by political, social, and religious feuds of the most serious kind, such as Scotland then was, there should have arisen the first of those great thinkers who in the course of the seventeenth century brought Great Britain to the highest point of achievement in the domain of Mathematical Science.